I0555425

GOD'S
Eclectic
Arctic
ANIMALS

Mary Ann Winslow

God's Eclectic Arctic Animals!

Book 8 of the God's Cool Creation Book Series

Copyright© Mary Ann Winslow 2023

All rights reserved. No part of this book may be reproduced or transmitted in any form or by any means without written permission from the author.

ISBN: 9798218251987

Author and Artist - Mary Ann Winslow, PhD

Cool Creation Press
Prescott, Arizona
coolcreationpress@gmail.com
www.godscoolcreation.com

Disclaimer: This book contains general science information intended to help the reader better understand basic principles. It is understood that some specific details have been omitted.

Dedicated to all you nature-loving kids! God created you in a very special way and loves you so much!

John 14: 21 And he who loves Me will be loved by My Father, and I will love him.

Where is the coldest, most barren place on Earth? The Arctic! However, God created in animals the most amazing ways to survive and thrive! Let's find out more about this incredible place, and about how animals use God's specific designs to live there.

The Brooks Range spans 700 miles east to west!

No thanks! Not this animal!

It stretches across Alaska and Canada.

The Alaskan Arctic has high mountains, coastal plains, and the tundra - treeless plains with frozen ground called permafrost. The Brooks Range, the northernmost part of the Rocky Mountains, is an alpine tundra, or a tundra in the mountains!

Where is the Arctic? It's located 66.5° north of the Equator - the top of the world! It spans eight countries - Canada, Greenland, Iceland, Norway, Sweden, Finland and the United States. And you guessed it! It's cold! The average lows in the Alaskan Arctic are about -40°F!

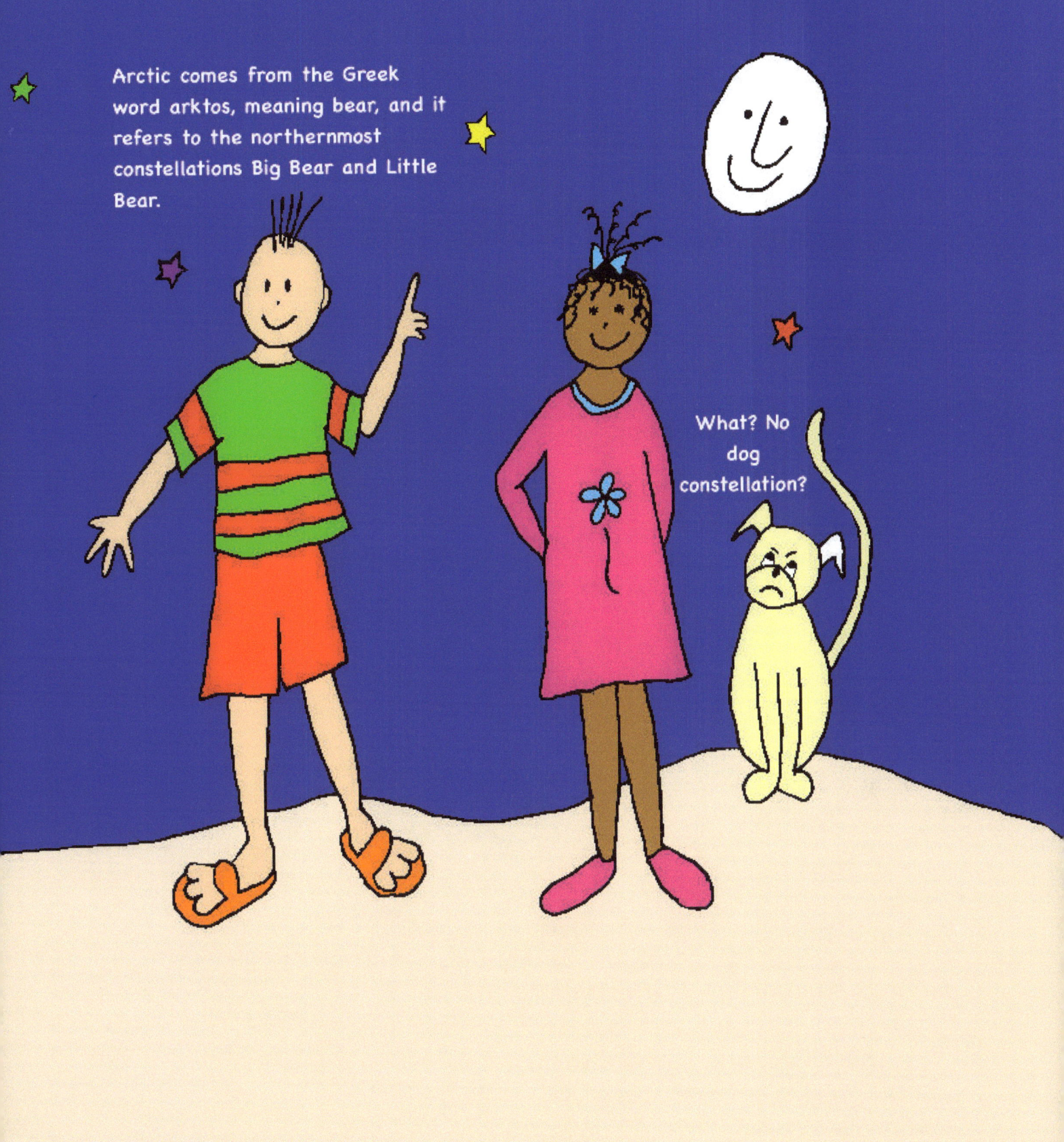

23.5° TILT

60N
30N
EQUATOR 0
30S
60S

Why so cold? Because the 23.5° tilt of the Earth is pointed away from the Sun in the winter. This means that it stays dark all day! When the Earth's tilt is pointed towards the Alaskan Arctic, it stays light all day and all night!

During the summer, the top layer of the permafrost melts, which is great for insects and animal life! The insects breed, or multiply, due to the shallow water created by the melting permafrost.

The birds then feast on those insects, and other Arctic animals feed on them! What a perfect food web that the Lord has created!

Can people really live in the Arctic? Actually, about four million people over those eight countries do. In the Alaskan Arctic, about seven thousand people live there.

We were here first!

The majority of people are the Inupiat, who lived in the Alaskan Arctic centuries before gold was discovered in 1898 and others came!

Not much plant life grows in the Arctic due to the short growing season, frigid temperatures in the dark winter, and strong winds. There must not be many animals, then, that live in the Arctic, with such harsh temperatures and little vegetation. Right?

Wrong! There are polar bears, wolves, foxes, hares, owls, lemmings, grizzlies, whales, orcas, narwhals, seals, sea lions, walruses,* musk oxen, eagles, many species of migratory birds, Dall sheep, caribou, and fish, to name a few!

*Read God's Sassy Sea Mammals for more information about polar bears, seals, sea lions, walruses, narwhals, orcas and whales!

The Lord provides many creative ways for animals to survive! Some animals, like the musk oxen, Dall sheep, and caribou, have an extra layer of special fur to keep them warm; others have a thick layer of fat called blubber, like the seals, whales, and walruses.

Polar bears and caribou have special hair that is hollow and traps air for insulation. This unique hair helps them float when swimming! Polar bears have black skin that absorbs the Sun's rays, which also helps to keep them warm.

Some animals' fur, like the Arctic fox, Arctic hare, and baby harp seal, change colors to match the seasons. This acts as a camouflage (KAM oh flahj) to protect them from predators.

We're special!

The fish that live under the ice have a special type of blood that acts like the antifreeze your parents put in their cars in winter!

The musk ox is a five foot tall, 800 pound stocky animal with an outer coat that almost reaches to the ground! Underneath these outer hairs is a thick wool called qiviut, which they shed in the summer. Qiviut is eight times warmer than sheep's wool! One of the lightest, warmest wools there is, it's used to make sweaters, scarfs, and mittens.

Musk oxen are passive animals, but when threatened by a predator such as a wolf, they make a circle around their babies, with their horns facing out. These horns are two feet long, so the wolves take off pretty quickly!

Where to?

The large caribou herds migrate to different parts of the Arctic to find food. They give birth on the coast, then head north to escape mosquitoes! About 490,000 caribou from the Western Arctic herd alone travel almost 3,000 miles each year!

God gave the caribou special feet for traveling! They spread apart, which helps them walk on the snow and the tundra. They're also amazing swimmers! Those hollow hairs help them float, allowing them to cover a lot of "ground."

I'm resting!
I'm no sheep!

Dall sheep live in the northernmost part of the Alaskan Arctic - the Brooks Range. They live in this steep and rugged ground to have their babies, and to escape predators, such as the brown bear and wolf.

God gave them an undercoat of fine wool to keep them warm. The guard hairs on the outside are long and hollow, "guarding" them from the harsh winter weather. On their horns are growth rings called annuli (ANN yoo lee). By counting the rings, you can tell the sheep's age, just like a tree!

Weaklings from the lower 48!

The Arctic wolf has shorter legs and smaller ears than the gray wolf; this helps to keep them warm in the frigid winter. They also have different layers of fur - the inner layer provides a waterproof barrier, and the thick outer layer protects them from the freezing cold.

No, they don't howl at the moon! They're just "talking" to other wolves! They're very social, living in packs of five to seven or more, and communicate by making various sounds and tail movements. They're fast, too! They can run 40 miles an hour, and because of their sharp sense of smell, can detect other animals as far as a mile away!

The brown bear, or grizzly, has long claws and sharp, long teeth, and weigh up to 700 pounds! When standing, they're eight feet tall, and can run 35 miles an hour! Those long claws are used to dig their dens for the winter months of hibernation, but it's not a deep sleep and they're easily disturbed.

Get this! They can stay in their dens for up to seven months, don't eat, and rarely go to the bathroom, if at all, during this time! Their babies are born during hibernation in the protection of the den, and their moms care for them for two years or more.

Birds migrate thousands of miles to get to the Arctic, but the all-time great migrant is the Arctic tern! It flies from North Pole to South Pole and back every year - over 22,000 miles! Some scientists have found that it's actually 55,000 miles, because they follow wind patterns instead of flying a direct route! They nest on the ground and will dive bomb any animal repeatedly, no matter how large, that gets near its nest. Watch your head, polar bears! Ouch!

How amazing are the Lord's designs for all of His animals, even in the Arctic!

Thank you, Lord!

www.ingramcontent.com/pod-product-compliance
Lightning Source LLC
Chambersburg PA
CBHW041446120626
46547CB00002B/365